POSTMODERN ENCOUNTERS

Ecology and the End of Postmodernity

George Myerson

Series editor: Richard Appignanesi

ICON BOOKS UK
TOTEM BOOKS USA

Published in the UK in 2001
by Icon Books Ltd., Grange Road,
Duxford, Cambridge CB2 4QF
E-mail: info@iconbooks.co.uk
www.iconbooks.co.uk

Published in the USA in 2001
by Totem Books
Inquiries to: Icon Books Ltd.,
Grange Road, Duxford,
Cambridge CB2 4QF, UK

Sold in the UK, Europe, South Africa
and Asia by Faber and Faber Ltd.,
3 Queen Square, London WC1N 3AU
or their agents

Distributed to the trade in the USA by
National Book Network Inc.,
4720 Boston Way, Lanham,
Maryland 20706

Distributed in the UK, Europe,
South Africa and Asia by
Macmillan Distribution Ltd.,
Houndmills, Basingstoke RG21 6XS

Distributed in Canada by
Penguin Books Canada,
10 Alcorn Avenue, Suite 300,
Toronto, Ontario M4V 3B2

Published in Australia in 2001
by Allen & Unwin Pty. Ltd.,
83 Alexander Street,
Crows Nest, NSW 2065

Series editor: Richard Appignanesi

ISBN 1 84046 279 5

Typesetting by Wayzgoose

Printed and bound in the UK by
Cox & Wyman Ltd., Reading

For Yvonne
with love

Ecology: The End of the Modern World?

Ecology is often assumed to announce the end of modernity – the final breakdown of modern confidence, the settling up of accounts between exploited nature and an exploitative society. Ecology discloses the disasters caused by the excesses and limitations of the modern: pollution by two centuries of industrialism; over-population supported by technological advances and driving them in turn further forward; viruses and bacteria renewed by the medical science that was meant to destroy them. Meanwhile, another ecological 'lesson' is that factory farming has finally warped our relationship to what remains of the countryside: with the spread of BSE, foot-and-mouth disease, a jugful of viruses.

This direful ecology appears to confirm the more abstract message of postmodern theory: the 'end of modernity', in the famous phrase of Gianni Vattimo.[1] The rise of ecology seems to support the diagnosis of the postmodernists, notably Jean-François Lyotard, that modernity was a temporary phase of self-destructive over-confidence. There are many postmodern theories, but they share the view that modern culture was infected by a narrow rationalism, a naïve faith in progress and in Western techno-reason as the salvation of mankind. Ecology appears to be the news of its come-uppance.

This book stages an 'Encounter' between ecology and postmodern thought, an 'Encounter' which challenges the assumption that together they seal the fate of modernity. On the contrary, we shall see that ecology is not postmodern at all. The ecological vision, with all its disastrous news, belongs to a moment of 'modernisation', another modern leap towards the future. Ecology confounds many of the premises of postmodernism, and renews, in new ways, the grounds of modernity. In the year 2000, ecology announced not the death of modernity but the end of its shadow, *postmodernity*.

Ecology does tell a new story about the modern world; a story that alters our perception of science, technology and progress. But this new insight, this 'ecological enlightenment' in the phrase of Ulrich Beck,[2] is thoroughly modern. To define ecology as postmodern is to skip over the difficult relationship between ecology and the modern outlook, a relationship that will increasingly shape our society's future.

This 'Encounter' gives a glimpse of how ecology acts as a modernising influence, in two contrasting ways:

1. Legitimating the mainstream

Ecology, the new understanding of man and nature, can become the newly legitimate face of mainstream modernity. With all its bad news, ecology can be used to declare a 'great leap forward' (to adapt Mao) of the modern order, which now claims to add a grasp of environmental consequences to the previous industrial system. This ecology is not post-industrial, but the herald of a new industrial future. For the mainstream authority of modern society, ecology can be a key resource. I shall call this role of ecology 'Ecological Relegitimation'. In the year 2000, ecology provided plenty of new arguments for the *legitimacy* of modern order.

2. Radical ecology

There is also a different potential to the new ecology. Modernity has always had within it a critical element, a dissident side. Although ecology can be adapted to support the mainstream, at the same time this ecology is the source of an alternative vision. This alternative is not postmodern. On the contrary, it is a new wave of modernism, a new radical modernism, the successor to the theories with which the twentieth century began. This new modernist critique expresses itself differently: it is diffused, has no central work or author, and speaks from many places. But this radical ecology is also present within the mainstream, and has the potential to reshape the development to come. In homage to earlier modernism, I shall refer to this radical ecology as 'The Ecopathology of Everyday Life'.

My overall argument will converge with that of Ulrich Beck, when he insists that ecology belongs to 'the waves of modernisation that face us'.[3] For better and worse, the autumn of 2000 shows how ecology takes us deeper *into* modernity, and not beyond it. The year 2000 brought home the news of ecology, and at the turn of 2001, we stopped being postmoderns.

The Appeal of Postmodernity

The Postmodern Condition

To set up the 'Encounter', we need to ask: what is meant by 'postmodern'? Diversity is a major theme of postmodern thought, and so there is more than one postmodernity. But they all originate in a perception that modern reason has reached its limits, both philosophically and in practice. This founding premise was proclaimed most forcefully by Lyotard, beginning with *The Postmodern Condition* (1979). In the 1980s, Lyotard enriched his diagnosis with the most original and lasting postmodern concept, *The Differend*. In this phase, there was a rich symbiosis between philosophy and fiction. The great novels of writers like Milan Kundera and Salman Rushdie carried forward the postmodern vision. In the anti-sociology of Jean Baudrillard, the apocalyptic streak of the theory took vivid hold, notably in his fragmentary masterwork – *Cool Memories*.

To many of us, there seemed something intuitively right about the postmodern view, the theory that modern society and thought had come upon their limits. Even if you did not take the postmodern theories literally, they seemed to catch hold of something that had altered. Postmodernism became a powerful symbiosis of theory, fiction and news. The high tide was the years immediately after 1989, when the Berlin wall came down, and it seemed as if we had left modern history behind. Postmodernity marched with increasing confidence into the early 90s, and its influence spread far beyond the writers and thinkers who explicitly took its name. For example, the most influential 'postmodern' argument was Francis Fukuyama's *The End of History*, which did not use the term. Jacques Derrida avoided the label 'postmodernist', yet his work became a classic expression of the postmodern blend of fictional techniques and philosophical premises. At the heart of a Derrida text like *Cinders*, there is the news that modernity has consumed itself, that only the toxic after-effects remain. This section extracts the core diagnosis from this postmodern outlook, so that we can then see how far events in 2001 confirm or displace it.

Lyotard based his diagnosis of modernity on the concept of 'Delegitimation'.[4] Modern society had not simply driven forward a wave of technological advances. It had found a new way to *justify* this perpetual change – economic and institutional, as well as technological change. A modern worldview had guaranteed the 'legitimacy' of the modern state, the modern economy and modern life. Lyotard attempted to define the modern worldview, in order to explain why the justification no longer worked. Increasingly, he argued, the citizens of self-proclaimed 'modern' society were not convinced by the outlook that was meant to justify their world to them. He called this process 'Delegitimation': we could not take seriously the underlying vision of our own way of life. Confronted by many of the changes of the previous decades, people in France or the USA or Britain had felt that they were living through the turmoil of progress, and that their lives were investments in that future. The 'postmodern condition' begins when that justification fails.

There were many symptoms of this 'Delegitimation' of modernity. The most famous was what Lyotard called 'The End of Grand Narrative'.[5] Modern society had justified itself, in the thoughts of its citizens, by recycling itself as a story of progress. In this 'grand narrative', several advances went together. There was the onward movement of science towards the truth, and alongside, as an inevitable accompaniment, there was the march of the Western model of democracy. The scientific plot Lyotard called the 'narrative of speculation'; the democratic plot was the 'narrative of emancipation'. Together, they told of the universal future of mankind.

Lyotard pointed out a paradox. Science – he presumed – was not meant to tell stories: it was supposed to be about arguments, theories and experiments. But this same science, he held, could only justify itself by subscribing to this grand narrative. He called this twist the 'return of the narrative in the nonnarrative'. Here Lyotard gave a vivid example, one that is important to the story of the year 2000. We can catch him switching on the television and gazing with increasing annoyance as an expert is interviewed about some pressing problem, or some new discovery:

[W]*hat do scientists do when they appear on television or are interviewed in the newspapers after making a 'discovery'? They recount an epic of knowledge.*[6]

Through the media, science was telling stories of heroic speculation, the wave of breakthroughs that were the engine of social and political progress. All these little stories added together into the grand narrative, or, as Kundera called it, 'The Grand March'.

'Delegitimation' begins when this grand narrative stops convincing everyone. Nobody can find the route map for the next lap of the Grand March.[7] Of course, many people had doubts all along. But now no one really has the old feel for progress: 'The grand narrative has lost its credibility.'[8] We have, in this diagnosis, lost more than a good story. We are bereft of a common world, without 'the unifying and legitimating power of the grand narratives of speculation and emancipation'. The modern world was both held together, and justified, by an ongoing narrative. On the other side, there is a world that is neither unified nor 'legitimated'.

As the narrative lost its spell, other changes were meant to occur. Postmodernity would bring the end of universal 'metalanguage' – Lyotard's term for the largest perspective, the one that fitted together all the elements of the modern story. This metalanguage was able to reconcile physics and economics, politics and sexuality. Progress was the 'metalanguage' of the modern world, and, behind progress, there was the triumph of science. In the postmodern era, according to Lyotard, 'science plays its own game; it is incapable of legitimating other language games'. There is a language game of political argument, and another game of economic policy. Each has its own rules. But in the height of the modern period, those games also belonged to the larger game of scientific and political advancement. Now each game has separate rules. Science cannot be used to justify any other decisions, to win any other arguments. In fact, science has enough trouble justifying itself: 'above all, it is incapable of legitimating itself, as speculation assumed it could.'[9]

Postmodernity would also thrive on the new Information Technology, or, as Lyotard then called it, 'computerisation'. Of course, knowledge continues to expand – exponentially. There has never been so much new knowledge. But for postmodernists, like Lyotard and Baudrillard, there is a gap between this information and the old forms of knowledge. You can have as much new information as you like – it will not bring back the old certainties. In fact, it is partly the flood of information that has destroyed the certainties of the modern period:

Seen in this light, what we are approaching is not the end of knowledge – quite the contrary.[10]

In the modern period, new knowledge unified the world. Now new information breaks the world apart. 'Postmodern Science' is about 'such things as undecidables, the limits of precise control, conflicts characterised by incomplete information. It is changing the meaning of the word knowledge.'[11]

This postmodern theory of information was subtle, and it was capable of refined development, notably in Lyotard's *The Differend*, which discussed the postmodern future of ethics. Before we look at those developments, we need to see how Lyotard created his concept of 'Delegitimation' in response to a contrary view. In the preceding years, the German social theorist Jürgen Habermas had proposed a concept of 'legitimation crisis' to try to explain what had changed about Western politics and culture after the 60s. Lyotard formulated his 'postmodern condition' by reversing this previous theory in various ways. For example, Habermas had proposed that societies are legitimate when they have rationally valid arguments, or arguments that are valid in the terms of their time:

Legitimacy means that there are good arguments for a political order's claim to be recognised as right and just ...[12]

Lyotard replaced 'good argument' with 'grand narrative' in order to develop his Delegitimation theory. Habermas too was suggesting that something had changed. He argued that in contemporary society we had reached a phase where 'it is realistic to speak today of legitimation as a permanent problem'. There is never a settled state of legitimacy any more, and at critical moments deep conflict breaks out: 'Such conflicts can lead to a temporary withdrawal of legitimation.'

These temporary disturbances Habermas called 'legitimation crises'. But he remained convinced that there were always more good arguments to be found. Modern society has a strong 'legitimation potential'. Habermas could not see modern Western society as permanently at a loss: '[T]he grounds or reasons ... can be mobilised'.

Lyotard's 'postmodern condition' is essentially conceived by regarding the legitimation crisis *insoluble*. Postmodernity is an incurable legitimation crisis, one that ends the whole game of legitimation. For Habermas, such a state was, and remains in his later work, implausible. This is because he finds too great a rational potential in modern culture. For Habermas, after the most primitive societies, '[a]rguments took the place of narratives'. By modern times, we have developed a society that is, at least in theory, founded upon 'rational agreement'.[13] Habermas is a critic of modern society, not a conservative. But he bases his critique on the failures of the modern world to fulfil its true rational potential. Postmodernity, for Habermas, is a pessimistic denial of this rational potential, a surrender of the possibility of a truly legitimate political and social order. Ulrich Beck provides an alternative defence of the potential for modernisation against the postmodern concept – darker than Habermas's defence, as we will see, but also founded on the hope of a rationally defensible world.

The Rise of Postmodernity

In our account of the year 2000, the question will be: what has happened to 'legitimation' in our time? And in particular, what is ecology doing to the processes of legitimation and delegitimation? In the run-up to the millennium, there has been a phase during which Lyotard's version of a postmodern approach said rich things about its contemporary world. In the 80s and early 90s, he refined his personal analysis of Delegitimation and what it means in practice. In *The Differend*, he proposed that we had entered a world of ethical impasses where '[o]ne side's legitimacy does not imply the other's lack of legitimacy'.[14]

There is, he insisted, no longer a universal rule of judgement. We are left with 'phrases' that belong to different games. Judgement is still possible, but only within those locally valid rules. There is no way to join all the games together. Postmodernity is 'heterogeneous'.[15] It resists the unifying pull of modern reason: 'There are as many universes as there are phrases.'[16] Another key term for this state is 'Incommensurability', the impossibility of applying a common standard of measurement to our different decisions and dilemmas.[17]

There did – and still does – seem to be a lot of

intuitive appeal to these theories. Baudrillard summed up this appeal to intuition in a postmodern poetic slogan:

Technology evolves, language changes, the voice breaks, fate overtakes us.[18]

Every day brings its new dilemma, often arising out of new information or new technology, and it is hard to imagine approaching them all consistently. For Lyotard, and his many followers, the old game is up: there can be no grand narrative any more, no unifying 'metalanguage' for all our problems, no *modern society* in the familiar sense. We are left in our fragmentary universes, abandoned and perhaps liberated.

Lyotard and the High Tide of Postmodernism

At the height of the postmodern tide, Lyotard told a 'postmodern fable' to replace the modern grand narrative of scientific, and political, progress. This fable was, in fact, elaborately scientific itself. The point was that no scientific story could have a human or humane meaning. Science still tells stories, but there is nothing we can do with them. We cannot use them, in particular, either to justify or to criticise our way of life, our political order.

Lyotard's 'A Postmodern Fable' (1993) begins with the end of the world, seen with dispassionate objectivity:

The sun is going to explode ... The end of history has already been foreseen ... The lifetime of a star can be determined scientifically.[19]

The theme of the fable is that there is only 'a finite quantity of energy'.[20] The fable is not really about humans at all, so what human significance can it have?

The hero of the fable is not the human species, but energy.[21]

Lyotard states clearly that his fable is counterpoised to modernity and its grand narrative: 'The fable ... is postmodern.'[22] He wants his story to be an antidote to 'the contagion of modernity'.[23]

By now, Lyotard had generalised his concepts of modernity and postmodernity. He had traced the infection of modernity back to 'early Christianity' which remade classical history into a story with the moral: if you have faith, then history will bring you grounds for hope. Only the faithless need despair. From this original spin, modernity then derived its delusive grand narrative: 'Over countless episodes, lay modernity maintains this temporal device, that of a "great narrative"'.[24] But Lyotard insisted that:

The postmodern fable tells something completely different.[25]

But does the fable illuminate our world? Is postmodernism playing *our* tune?

Autumn 2000: The Millennium Finally Arrives

Millennium on the Forecourts: The True Countdown

No new age began on the first day of January 2000. Speeches were made about new eras, and futures, but nothing very different appeared, not even the millennium bug. Then, in the UK, there was, abruptly, a millennial autumn. In fact, it is possible to date the millennial moment as 13 September 2000.

On Thursday, 7 September, UK petrol was due to rise 2p in price. There had been a mounting controversy about this situation. A hundred farmers and lorry drivers staged a protest and tried to block an oil refinery in Cheshire. The action was based on protests in France in August – a move which certainly fits with Baudrillard's postmodern theory, where there is never an original action. This protest is only, in these terms, an echo of previous media images.

This was all minor stuff. On Friday, 8 September, lorries began a 'go-slow' and more protestors blocked a refinery in Pembroke. Suddenly, the momentum gathered. By Saturday, 9 September, no tankers were leaving most refineries around the UK. On Sunday, 10 September, there were widespread reports of the 'panic buying' of petrol. The weekday routine was under threat: the millennium had arrived at last.

From here the 'crisis' accelerated. On Monday, 11 September, emergency powers were declared, a prelude to bringing in the army to move petrol supplies. The crucial factor was that all opinion polls and phone-ins revealed 'massive' public support for the protest. It seems that the blockade was being undertaken on behalf of all the other drivers who were going to have to pay those impossible prices, the highest in the 'developed' world. Now there followed two memorable days that have left images in the collective memory. First, Prime Minister Tony Blair promised that 'normality' would be 'restored' within 24 hours. This was on Tuesday, 12 September. Then there was Wednesday 13th, my choice for Millennium Day: over 90 per cent of petrol stations were out of supplies. There was food rationing and panic buying. The roads fell silent. At last, the old world had passed, as millennial predictions had always insisted it would. Everyday life had, it seemed, ended. Yet the next day, it was all over. On Thursday, 14 September, most of the protestors went home and the blockades were broken. More action was promised, but the moment never returned.

This story has some very postmodern features. There is the repetition of previous news. There is the sharp change of gear in which a minor item becomes history and then fades away abruptly. This feeling of 'crisis' and 'anti-climax' fits with the story-telling approach of, say, Milan Kundera: the trivial is momentarily elevated, and then abandoned once more. But above all, Lyotard's concept of 'Delegitimation' seems to make sense of the inner logic of the episode.

Consider some of the individual moments. On Saturday, 9 September, BBC News Online announces that 'Protests trigger fuel shortages'. Everyday life comes into an eerie close-up, in ways that certainly reproduce Baudrillard's fragmentary glimpses of mundane apocalypse in his *Cool Memories*:

Tesco in Bidston Moss on the Wirral reported it had run out of petrol while cars queuing for petrol at a Tesco in Allerton were causing traffic problems.

(BBC News Online)

This is just the everyday world. But coiled inside these scenes is an apocalyptic potential: it is as if ordinary glimpses are coloured in a strange light.

On Sunday, 10 September, the BBC carried statements by the 'People's Fuel Lobby':

We are doing this for the people of Great Britain.

(BBC News Online)

This was the beginning of a deeper challenge, one that did raise the question of legitimacy. For a few days, there seemed to be a conflict of the nation versus the state, with public opinion on the side of the protest. Was this to be the passive revolution of Baudrillard's silent majorities – revolution by phone-in, led not by orators but by pollsters?

On Wednesday, 13 September, the BBC reported that 'locals' were rallying to the support of the petrol blockade, as the police and army threatened to force the tankers through:

Hundreds of local people joined the protestors to lodge an almost unanimous vote in favour of continuing the protest.

(BBC News Online)

This was outside a refinery in Scotland where the new First Minister, Donald Dewar, declared:

I will not accept that government policy should be dictated by protests of this kind, which are so badly affecting our everyday lives.

(BBC News Online)

This was becoming a conflict between two kinds of legitimation: precisely the kind of impasse that Lyotard leads us to anticipate once the grand narrative has faded, once there is no universal criterion. We face competing versions of legitimacy. On one side, there is direct democracy: a count of hands on the scene, backed up by public opinion in various mediated forms. On the other side, there is formal democracy, the elected government. It is as if each has its own sphere. How can anyone decide between them?

Then, as the reports mount up, we do seem to enter a moment when the legitimacy of the authorities is in question:

Britain grinds to a halt as Blair's pleas are ignored
(*The Guardian*, 14 September 2000)

Everyday life is ending, with Britain 'close to shutdown'. Tony Blair is in no doubt that the issue is Delegitimation. Lyotard might have relished the use of the word 'credibility' in his statement:

At a Downing Street press conference, Mr Blair said; 'No government, indeed no country can retain credibility in its democratic process or its economic policy-making were it to give in to such protests. Real damage is being done to real people.'
(*The Guardian*, 14 September 2000)

Suddenly, the world seems not to respond to the call of authority. Clearly this cannot have arisen out of nothing, in two days. The episode does suggest that there is nothing very deeply rooted in this political order. It does not take much for our lack of faith to surface.

The protest breaks up, as arbitrarily as it came, on Thursday, 14 September. At this penultimate moment, BBC News Online declares that this is 'Environmental Gridlock'. Now the conflict is clearly defined as politicians versus people, state against society:

While politicians like to blame the massive rise in world oil prices for this latest fuel crisis, taxation is the real reason why petrol and diesel are so expensive.

(BBC News Online, 24 September 2000)

Behind the crisis, there is a policy:

Fuel prices are deliberately kept high as a means of fighting the very complex issue of global climate change.

(BBC News Online, 24 September 2000)

This moment brings into sharper focus the issue of 'Delegitimation'. There is a *scientific* theory – global climate change – and governments have decided to take it seriously – no doubt with other motives as well. If you don't believe the science (that global climate change is due to human activity in part), then you won't accept the legitimacy of the political authority (fuel taxation). On the other side, if you believe the science, you accept the political authority of those governments. These protests are a double challenge: to official democratic authority and to scientific theory. In the face of these blockades, the narrative of scientific progress goes with the legitimation of democratic governments and policies. The government draws its authority directly from the science. A society that no longer believes in scientific progress will also not accept the political authority of the government. And it looked, at least on 13 September 2000, as if we had decided to accept Lyotard's theory of Delegitimation!

It seems that the postmodern condition had made the news. One could not ask for a better illustration of Lyotard's theory – embellished by plenty of Baudrillardesque detail and looking eerily like an episode from a novel by Kundera. Even the withdrawal of the blockade fits: it has the incoherence of a postmodern crisis, rather than the logic of a modern problem.

The Autumn Floods: Science Coming True

But then something else altered the story. Events intervene which change the meaning of what went before. First, the weather turns rainy, and everyday life – which has just 'returned to normal' – begins to collapse for a different reason. There are local floods. On 28 September 2000, BBC News Online announces:

Warming climate 'means worse weather'

Next to the text on the web-page is a picture headed 'Snow in Jerusalem'. The substance is a scientific report:

A report by the World Wide Fund for Nature (WWF) says global warming is happening, and is probably already affecting the weather.

The report says that human activities are 'at least partly responsible', and car use is singled out in the surrounding discussion. The world faces '[m]ore extreme weather', and scientists claim that they can both predict and explain this process:

We conclude with reasonable confidence that we

are now experiencing the first effects of the increase of greenhouse gases in the atmosphere.

Lyotard foresaw a world where 'science plays its own game; it is incapable of legitimating other language games. The game of prescription, for example, escapes it.'[26] During the September crisis, this postmodern condition seemed close at hand. But now the theory begins to feel out of touch. Already, science does appear able to intervene – at least – in other language games, from politics to everyday life. In particular, science has not lost its link with the game which Lyotard calls 'prescription', the sphere of policies and remedies (green taxes etc.).

By science, Lyotard means specialised hypotheses, which are tested by remote experiments and recorded in alien idioms: the law of entropy is his classic case. Ecology – the science that is now impacting on our story – does not fit his conception of modern science, nor does it conform, as we will now see, to his brief sketch of 'postmodern science' which is limited to a realm of 'undecidables' and specialised controversies.[27]

The modernist poet and critic T.S. Eliot defined metaphysical and modernist art as 'the yoking together' of elements that normally hang apart. On

Thursday, 12 October 2000, the BBC News provided a great example of this modernist poetic in action:

UK will get wetter, experts warn

(BBC News Online)

On the one hand, we have the long-term theory, science at its most comprehensive: the fate of the earth can be foreseen. Compare Lyotard's postmodern fable of the entropic planet: a planet remote from everyday life, indifferent to human actions, and with no place for human agents. In this news broadcast, the story of the earth is woven together with today's bad rain, with your use of the car to go shopping, with the prices of petrol and the policies of the chancellor.

The online version of the bulletin shows a scene of a fire engine standing in a flooded street. Underneath, the commentary adds that this is '[a] sight destined to become more common in the UK'. Headline by headline, the everyday is being absorbed by the theoretical; the routine by the newsworthy; the perceptible by the abstract. This is a point of synthesis, whereas postmodernity envisaged continuous fragmentation.

Now, the petrol crisis reads differently. Instead of

the fuel blockade, we have the flooded road, and suddenly it seems there is a reason for the fuel taxes. Ecological science is tightening its grip on the news agenda. That was no ordinary rain:

Weather experts are warning that the UK should brace itself as this week's severe flooding will become more frequent.

The heavy rainfall causing chaos and flooding homes in the south-east may be linked to global climate change.

(BBC News Online, 12 October 2000)

Fittingly, the petrol blockade took place in bright sunshine. Now we have the answering downpour. If you were a zealous narrative interpreter, or a structuralist, you could contrast the missing liquid of the petrol crisis with the excess of this next crisis! The story is both rich and coherent. One of the main reasons why postmodernism isn't working is that Lyotard and others underestimated the story-telling powers of the modern world. In Lyotard's theory – and elaborations by others like Baudrillard – there was an assumption that grand narrative was also a bad story. That was one unspoken reason why we would be glad to see the back of it. We could then have more entertaining stories, like those in Kundera's novels. In fact, grand narrative always was a much better story than Lyotard realised. It was flexible, capable of modulation and irony, open to different circumstances. This news episode shows just how good modernity still is at telling its story.

The story is unmistakably 'grand': 'The number of days of very heavy rain could increase substantially', announces the Met Office. In such sentences, there is a fusion of the everyday weather forecast with the prediction of global climate change. In Lyotard's theory, there is no more market for a 'grand narrative' of speculation. That is, nobody will buy or sell stories of the advance of science leading to the advancement of society. But in this flood crisis, there *is* here a new kind of grand narrative of speculation. The theory confirms itself in front of our eyes. What you are experiencing is not just rain, but the proof of a scientific hypothesis. Thus your experience turns into a witness for the advance of science. We are all in the laboratory now.

True, the scene all around us is a disaster – but it is also a confirmation of progress, the success of a modern theory. That knowledge will, in its turn, legitimate remedies, as we will see. Such moments are the reverse of Lyotard's world of 'differend', of incommensurable games. These are points of linkage where ecology starts to become a new metalanguage, a new commensurability – and a *common measure*.

The news story also has a peculiar structure over time. Day by day, the everyday world is consumed by emergencies and incidents:

Floods cause chaos

(BBC News Online, 12 October 2000)

Again, this is a 'bad news' item that follows directly from the turmoil of the petrol blockades. This period is all about lack of mobility, the collapse of everyday freedom of access:

Some of the worst flooding in decades has caused chaos and widespread damage in southern England.

(BBC News Online, 12 October 2000)

You look out of the window, and you see the rain falling, the everyday disappearing:

Downpours on already saturated land and overflowing rivers flooded homes and businesses, cut off roads and delayed or stopped rail services.

(BBC News Online, 12 October 2000)

However, at another level, as we have seen, these are also moments of triumphant confirmation, of theoretical success.

The scenes are also gripping. This is news that produces episodes of excitement out of the everyday world. You could call this reportage 'The Serialisation of Everyday Life'. These rescues and floods are no postmodern fragments; on the contrary, each incident belongs to a bigger story, the biggest story of all – a global chronicle. This serialisation is the twenty-first-century form of grand narrative; after all, why would our story share the narrative structure of *Middlemarch*?

Ecology brings with it a new way of belonging to the grand narrative:

Rescue workers evacuated residents…

[A] *lifeboat crew rescue 20 men, women and children trapped in a Somerfield supermarket …*

Vernon Jay, a jeweller from the town [was rescued] … *attempting to reach his shop … caught up in the current …*

(BBC News Online, 12 October 2000)

Through the episodes, one term begins to stand out: 'risk'. For example, the Environment Agency 'warned people "not to take risks"'. Risk is also a good term for the wider story, in which modern society takes calculated gambles on 'getting away with' its ecologically damaging lifestyle. Risk is what makes these incidents from everyday life newsworthy. It is also the theme of the larger analysis. This is the world that Ulrich Beck called 'Risk Society'.[28] As we are seeing in these examples, Risk Society is culturally modern, not postmodern. The mindset is all about the effects of rational calculation, the impacts of technology; and the many failings or crises imply remedies drawing upon more *accurate* reasoning and *better* calculations. Ecology produces a grand narrative for a twenty-first-century Risk Society.

In October 2000, you can actually feel this grand narrative of ecology taking hold of everyday life. As we saw earlier, Lyotard's postmodernity was defined by the failure of the 'grand narratives' to unify and justify a modern worldview. But here we see a new unifying power – a narrative of the advancement of science that takes hold of everyday life. It is also a narrative that (re)legitimates the political authority of the government against the preceding protests. No wonder the opinion polls swung violently against the attempts to remount the blockades or go-slows! This episode can be understood as an example of the 'Ecological Relegitimation' of modernity. The protests had seemed like a popular rebellion, or even revolution; increasingly, they appear to be reactionary – the actions of those who are trying to resist the onward movement of the modernising world. Those heroic blockaders now seem like smallholders against progress, as in previous grand narratives going back to the nineteenth century: peasants against the march of history. These are small folk who simply have not understood the new language.

Lyotard talks of 'little narratives' as the remaining field for science.[29] But there could be no more grand narrative than global climate change – a narrative in the old grand style, where science legitimates both itself and political authority at the same time as telling a universal story. Here science actually reveals our entire story. Compare this autumnal millennium to Lyotard's postmodern fable – where the story is all about the system being *unresponsive* to human actions! In our world, the story is about how the world responds to our actions. Our lives fill with problematic meanings, the very reverse of

Lyotard's postmodern fable where meaning floods away from our everyday lives under the indifferent gaze of science.

By Thursday, 12 October 2000, the BBC is announcing:

Worse floods ahead as world warms

(BBC News Online)

In such headings, there is no distinction between the history of the world and today's muddied streets. The news speaks equally a language of global prediction and everyday detail:

But what has happened recently could soon become relatively commonplace, if it is an early sign of climate change.

And not only environmental campaigners believe that it is.

(BBC News Online, 12 October 2000)

Activists and insurers, scientists and politicians, they all speak this common language:

The ABI [Association of British Insurers] *says: '... there is a reasonable consensus emerging that we are in for a period of much more extreme weather ... You can say the floods in the UK and southern Europe bear the fingerprint of climate change ... The latest work of the Intergovernmental Panel on Climate Change confirms that the climate is having an effect on the insurance industry.'*

(BBC News Online, 12 October 2000)

So while this is a bad news item, its language is consistently upbeat about knowledge: 'reasonable consensus emerging', 'the latest', 'confirms'. The most striking phrase of all comes from the insurance expert, as Beck might predict: 'bear the fingerprint'. Amidst the chaos, we are present at the solving of a detective mystery, a crime story. This is a moment in which the detectives have cracked the case. Now we know whodunit. (You did.)

In John Cleese's film *Clockwise*, the bedraggled hero creeps out of another ditch on his endlessly impeded journey, and utters the protest of a true citizen of modernity: 'It's not the despair I mind, it's the hope!' It is not the pessimism that takes this autumn out of the reach of postmodernity, but the optimism – about knowledge itself. This is, in retrospect, the moment when science relegitimates both itself and the modern approach – finding global visions and solutions. Ecology is the new grand narrative insofar as it has this universal range and reach. It is also all about coming up with answers.

Now if you put this news next to Lyotard's view of 'information', you can feel postmodernity slipping into the past:

> [W]*hat we are approaching is not the end of knowledge ... Data banks are the Encyclopaedia of tomorrow. They transcend the capacity of their user. They are 'nature' for postmodern man.*[30]

Lyotard argued that we would have data beyond any unifying vision or single theory. The data would not tell a story, it would remain fragmentary, a list of facts. But here is an extremely powerful theory, which synthesises a mass of data into a single story – and in that story sweeps up the everyday and the historic, the diagnostic and the prescriptive. Global climate change is clearly a move well beyond Lyotard's concept of postmodern information. We are returning, in a new way, to the unified narrative of modernity.

The BSE Verdict: The Dark Legitimacy of the Modern Order

At the end of October 2000, another ecological story came together. Into the midst of the climate story comes a moment of tragedy and scientific confirmation. On Thursday, 26 October, we had the publication of the Phillips Report on the crisis of BSE infecting cattle, and especially the Conservative Government's handling of the episode in 1996 when the 'risk' to humans was finally admitted. On Saturday, 28 October 2000, the follow-up news began:

BSE teams rethink size of epidemic

(*The Guardian*)

Immediately after the retrospective Report, there came 'confirmation that a 74 year old man has died of the disease'. BSE had been presumed unlikely to cross species barriers. The individual nightmare was also a scientific validation:

This followed the finding of Lord Phillips' BSE inquiry report, published on Thursday, that BSE might have existed in a series of rapidly increasing epidemic waves from 1970 until it finally became obvious as hundreds of cows began to die in 1987–88.

(*The Guardian*, 28 October 2000)

This news shares the structure of the climate change coverage. Things are worse, this is bad news; at the same time, a scientific theory is confirmed – within days of being announced. This is the dark relegitimation of the modern order of things. Only science can explain these frightening phenomena, even if the technological and industrial society has also produced them. Not only is science the source of explanations, and any possible remedies: it is also the necessary reference point for any legitimate political response.

The BSE coverage supports the relegitimation of modernity: science can synthesise the data; the mystery can be cracked. The detectives can still solve the crime, despite Baudrillard's insistence that at last postmodernity had perpetrated 'the perfect crime' and stolen reality itself from under our noses.[31] In the *Observer*, Sunday, 29 October, the BSE story is put together again. The Phillips Report shows how democracy and science must jointly overcome tradition and corruption: the enemies of science are also the corrupters of democracy. The

BSE story becomes a cautionary annex to the modern grand narrative.

According to the official verdict, the BSE crisis was made infinitely worse because the political authorities refused to follow their modern duty – they suppressed or impeded the science:

> *It was suggested that information should be passed around, the usual method when new scientific discoveries are made ... Plans to give information to universities and outside research bodies were blocked.*
>
> *If this theory* [of a new mode of viral transmission, conveyed to the Minister of Agriculture in 1988] *is correct we have to face up to the possibility that the disease could cross another species gap.*
>
> *This view would not be made public for another eight years. MAFF considered it too dangerous ...*
>
> (*The Observer*, 29 October 2000)

Science made rapid advances in the face of the possible disaster, but the authorities chose to be obstructive. The Report is a tale of the breakdown of the modern linkage between scientific and political legitimacy:

In August 1990 scientists successfully transferred BSE to a pig. In March 1993 the Lancet reported that the first dairy farmer had died of CJD, the human equivalent of BSE. That month Kenneth Calman, the chief medical officer, issued a statement saying that beef was safe to eat.

(*The Observer*, 29 October 2000)

So far, so bad. But the whole point is that the Report is an official verdict:

Eventually the weight of evidence overwhelmed the Government's defences.

(*The Observer*, 29 October 2000)

Relegitimation begins here: a rejection of the previous era, and a reaffirmation of the link between modern politics, democratic freedom of discussion and the progress of science. Phillips confirms the need for the new wave of political modernisation that brought the Blair Government to power.

There is a scientific discovery – the cross-species transfer of BSE – and soon a coherent theory follows. This theory takes hold on individual cases and has the potential to interpret our everyday world, illuminating new areas of risk. But the advance is suppressed by the very authorities that should have aided and disseminated it. In effect, those who impeded the science appear now as anti-democratic forces. As the BSE story comes together, the moral emerges. Science and democracy are allies. Together, they comprise modernisation, which now appears to be an endless process rather than an accomplished fact.

But, of course, the latest BSE discovery is bad news. The story is comprehensive, and hope waits on the horizon: in between, the world is darkened. This again is characteristic of the ecological version of grand narrative. Where BSE and climate change meet, in the millennial autumn news of 2000, we have the 'unifying power' of this ecological grand narrative – a unity as much tragic as encouraging.

The new grand narrative is a story in which science and democracy advance together, legitimising one another. Or they meet common obstructions and antagonists.

In this ecologised version, the grand narrative is about:

- The relationship of man to nature
- The responses of the world to our actions

- The cycles of pollution and cleansing
- Health, well-being and sickness

The new narrative is even wider in scope than any preceding story of human progress. But the concept of risk is spreading everywhere, and sometimes the warnings come too late. The narrative applies science to news, to politics, to everyday issues, using the concept of risk:

- These are episodes, not the fragments proposed by postmodernism
- The tension is suspense, not the 'indeterminacy' of postmodernism

We have seen how climate change turns protestors from a popular front into reactionary buffoons, blocking the path to modernisation. By early November, the news is that we are:

Living with climate change
(BBC News Online, 10 November 2000)

Modernisation has found its new agenda, a problem worthy of its powers:

Climate change ... is very hard to slow down.
(BBC News Online, 10 November 2000)

Everything is tied together. Our world is coherent, interlinked by this universal challenge:

... it is so complex, with impacts on one area having consequences on another thousands of miles away, or a long time into the future.
(BBC News Online, 10 November 2000)

Of course, the story is troubling. Our acts rebound obliquely and their effects are amplified at a distance. But look at the sheer scale of the narrative! Only a month ago, we were in a world of spontaneous uprisings. Now that 'peasants' revolt' seems like a minor diversion. Ecology has put the capital H back in 'History', which becomes again the story of heroic modernisation in the face of intractable difficulties bequeathed by the pre-modern past:

There is the option of doing things differently ... renewables like solar power, wind and wave energy, ... fuel cells and hydrogen-powered cars.
(BBC News Online, 10 November 2000)

There follows the international Hague conference on climate change. Agreement eludes us, for now – History is on hold. The moment has a message: there are limits to the 'Ecological Relegitimation' of the mainstream. The new consensus has still to materialise. The BBC (12 November 2000) features a British minister speaking for the future, as his international colleagues lag behind.

The British Environment Secretary, John Prescott, brandishes a sandbag from flooded York at the delegates: wake up to the new challenge! If he looks silly, rather than heroic, that is a sign that the narrative has not yet taken hold of everyone. Immediately after the conference, the BBC staged an online 'debate' ('Talking Point', BBC News Online, 4 December 2000). Although the voices differ and quarrel, they have a clear question on which to take sides. Here is a voice of modernising ecology, demanding a global clean-up:

Remember it is not just cars that emit greenhouse gases, it is also factories that produce packaging, air conditioners, farm equipment ...

Nigel, UK

Others give the line a moral twist:

You and I are just as guilty for not facing the issue as well …

Svenn, Denmark

On the opposing side, some deny the scientific basis of this new politics:

As for global warming, there isn't any scientific evidence that it's even happening.

Sam, Texas

Or, if they accept some of the science, these eco-sceptics contest the interpretation of the data:

Why do we assume that it is human progress that is creating an ever so slight increase in average earth temperature?

Kevin, USA

Others concede the facts, but denounce the remedies:

If the US is to take a 10 per cent reduction in greenhouse gasses it will damage our economy, and what is bad for us will be even worse for you.

David, USA

Of course, we don't all agree; but we are all playing the same game. The rules of the argument are clear and everyone is following them. If you want to challenge the new modernisation, you have to either deny or reinterpret the science. Lyotard envisaged a world where argument fragments, without any common rules. Instead, we live in a world where the arguments are inter-related, across issues (such as BSE, global warming), places and times. The winner is still to be decided: 'Sam, Texas' has brought President George W. Bush to power, but the Climate Conferences reconvene. In the terms of Jürgen Habermas, the new consensus is still lacking, but the story continues.

Radical Modernism: Small is Significant

Ecology is a powerful resource of mainstream modernisation. Against the protests of the (now) reactionary elements, democratic authority pursues its modernising way. Ecology is the science that presides over this new modernisation – the low-carbon age, the era of green taxes and new standards. In Habermasian terms, ecology can be mobilised to give good reasons for new modernising initiatives. Or you could draw – upside-down – on Lyotard's postmodern idiom and say that ecology weaves a grand narrative around the new politics, a new kind of story of progress. In episodes like the October floods, and the Phillips Report, mainstream modernity became – or tried to become – ecologised. The goal becomes not just a technologised world, but also an ecologised world.

But there is another aspect to the impact of ecology. Beyond the mainstream vision and project, there is a radical influence, a rational extremism. In fact, this radicalism, this analytical fanaticism, is present in glimpses inside the mainstream. The next section examines this other ecology – the places where ecological perspective leads to more thoroughgoing critique: 'The Ecopathology of Everyday Life.'

At the start of the twentieth century, Freud's Psychopathology ended the distinction between normal and pathological psychology in everyday life. He headed his most populist volume, *The Psychopathology of Everyday Life* to underline this challenge. By analogy, this new 'Ecopathology of Everyday Life' subverts the division between normal and pathological, from an ecological point of view. Freud saw little psychopathological symptoms everywhere in normal life. Ecopathology sees eco-symptoms breaking out everywhere in our mundane routines.

This Ecopathology is present within the mainstream media, but only sporadically and as if by accident. Ecopathology of the everyday always emerges at a particular kind of moment: when expertise reads deep significance into tiny and apparently insignificant details of our lives. Ecopathology, like Psychopathology, finds great significance precisely where everyone else sees none at all. The problem is that in seeing this extra meaning, Ecopathology pushes beyond the needs of the main argument, and threatens to alienate those who are happy enough with the ecological approach in general.

An example of this Ecopathological interpretation, this penchant for extreme analysis of minutiae, is a moment when the news paused over flooded Bognor Regis:

Homes in Bognor Regis had to be evacuated as drains failed to cope with standing water. The Environment Agency sent its emergency planning unit to the scene.

(BBC News Online, 10 October 2000)

The Ecopathology of Everyday Life begins when you insist that there is no such thing as simply a blocked drain. This blocked drain is a symptom of global climate change, a mundane confirmation of a deeper meaning that has been discovered behind everyday life.

At its most extreme, ecological interpretation gives a deepening significance precisely to the most trivial details. The effect can be disorienting. Whereas the main thrust of the ecological argument is that we should be more reasonable, these are moments beyond ordinary reasonableness. You never know when *your* blocked drain will also have this latent meaning. So the most radical moments in the mainstream news come when ecological experts make their almost involuntary stray remarks that reach beyond the immediate agenda. Almost every environmental news item has such a moment. For example, the BBC, on 12 November 2000, was previewing the Hague conference on climate change. We are hearing from a climate scientist. He does indeed have a grand narrative to tell: he presents himself as one voice of a new advance, comprehensive and politically urgent. But he also lets little remarks drop:

Things may grow better for a while and there has been a move towards outdoor living with barbecues and café society ... Quite a lot of it will be quite nasty.

(BBC News Online)

This is classic Ecopathology, a flicker of analytical extremism amidst the restrained persuasiveness. Suddenly, he takes petty details and reinterprets them to show their hidden significance. These cafés are hidden expressions of climate change, and so they are related to the floods that have swept away the ordinary world. As in Psychopathology, subterranean connections run between the most harmless and the most pathological.

You think you are just sitting outside a café. But the Ecopathological interpretation reveals the hidden meaning of your little habits – the latent consequence of your moment at that table. Your changing routine is a covert – and self-concealed – admission that you already know about climate change. Why should the flood surprise you?

The Ecopathology in these remarks is profoundly modernist. The modernist features of this style of interpretation include:

- The view that expertise is needed to find the true meaning of everyday details. Trivial experience has indirect significance, which you cannot 'get' through common sense, the direct view.
- The disturbing idea that we do not understand the effects of our own actions or intentions.
- The negative principle of self-deception: bad consequences are amplified because they are unacknowledged. We cannot control what we refuse to notice.

Like Psychopathological theory, Ecopathological theory expands its own central concept. If tables on the pavement, and blocked drains, have hidden meaning, then presumably so do all the other trivialities. It is precisely at that level that the big messages are conveyed, and ignored. Every little habit has an ecological meaning. After Freud, lots of little things counted as 'psychological' that used to be innocent; and now lots of other trivialities count as ecological. Ecopathology is the god of trivial things.

A hundred little asides slip into ecological discussion every time there is a relevant crisis. On the edge of the main story, in the margin of the central argument, there are always these over-vivid glimpses of really tiny details, the fine texture of lifestyle or landscape. These details always have a spontaneous energy, beyond the demands of the main purpose. What these little asides say is alarming: you have not really noticed how you are living, or where you are living. How different your tiny routines look when reinterpreted from an expert standpoint. These points of analysis are the radical – Beck would say, the 'alternative' – face of eco-expertise. This 'Ecopathology' is a radical potential for an *alternative* modernism – as radical as the Freudian interpretation at the start of the century. If we look back at Freud, we can see just how close the affinity is. What we are glimpsing here, in a thousand stray expert asides, is a revival of the radical energies that have always been part of modern reason.

Freud's *Psychopathology of Everyday Life* was the outburst of one of the great modernist theories – a total system capable of interpreting everything, a universal metalanguage. The book was also a self-declared step in the advancement of universal understanding. Psychoanalysis was, in addition, a form of prescription. So far, so modern. But this Freudian modernism was not a mainstream theory, and its take on the grand narrative was anything but supportive of the legitimacy of the established order. On the contrary, Freud's modernism was critical, alternative and dissenting, and a hundred years later, the Freudian outlook still disturbs the mainstream.

The parallel between Freud's Psychopathology and contemporary Ecopathology is close and consistent. Behind the analogy lie the five main principles of Radical Modernism:

1. Small is significant

Freud formulates this principle psychologically:

I applied psychological analysis to the frequent circumstances.[32]

We live amidst a clutter of tiny habits, mere circumstances, we think, that happen to accumulate while we are pursuing what really matters. Freudian Psychopathology reverses the priorities: small is significant, and tiny is even more significant. Circumstantial is essential. Ecopathology applies the same principle to a different dimension of everyday life. The café table, the blocked drain, the short car journey: the almost invisible circumstances of your life betray, to the expert eye only, your ecological significance.

2. There is always a deeper explanation

Here the Freudian version is that each detail of common experience

admits of an explanation much more far-reaching than that which the phenomenon is ordinarily made to yield.[33]

This principle is deeply provocative when applied psychologically, and even more provocative when applied ecologically. The tiniest details carry huge meanings: there is no escape from the big picture, the crisis follows you into your refuge.

3. To understand means to connect

Freud sets out systematically to show that everything in our mental life is 'connected in a discoverable way'.[34]

From this principle, followed through with radical extremism of the true modernist, there emerges a secret web that links the harmless surface of the ordinary world to a hidden meaning: the repressed crisis of the psyche, the hidden catastrophe of the planet.

4. The hidden meaning is always disturbing

Freud relishes those moments when his insights turn

nasty, when the interpretation recovers the darkness beneath the mundane surface:

But it was certainly surprising that the attempt to trace a harmless failure of memory back to its cause should have had to come up against matters in the subject's private life that were so remote and intimate, and that were cathected with such distressing affect.[35]

This principle is pursued with even more radical edge by Ecopathology. Whatever is hidden, and then revealed, will always be menacing. We would have noticed, if it had really been safe.

5. The less direct the connection, the more significant it will be

Freud is never more triumphant than when announcing that there is a sideways link leading outwards from a minor detail:

[I]*t can be characterised quite generally as an oblique relation.*[36]

Radical modernism gives priority to the indirect over the direct. The result is that what seemed important turns out to be insignificant: the real connections run outwards from the neglected nodes that you have overlooked. This style of interpretation is an important part of what Lyotard encouraged postmodernism to neglect. In its radical moments, modernism always did have a wicked sense of irony. Baudrillard seems to think that irony begins with postmodernism, after the literal-minded area of the modern. But the line of modernist irony runs straight from Psychopathology to Ecopathology.

The total effect of radical modernism, both psychological and ecological, is to suggest that there is far more meaning than we had ever supposed plausible – and this is because there are so many new connections: 'the multiplicity of the relations and meanings'.[37]

These new meanings connect mundane circumstances to the deep pathology that is threatening our lives.

> [T]*he slips of the tongue that we observe in normal people give an impression of being the preliminary stages* ... [that develop] *under pathological conditions.*[38]

This approach is modernist – in the terms that

Lyotard sees as displaced by postmodernity. There is certainly a grand narrative of speculation being expressed. The grand narrative is everywhere in Freud. His ideas are expressed as breakthroughs in the achievement of a new modern and scientific understanding. Psychoanalysis also belongs to the project of a universal metalanguage, which can embrace previously scattered spheres of experience. But in its original moment, Freudianism is also dissenting, embattled, radical, a counterforce to the mainstream authority. The same impulses are now reappearing under the sign of radical ecology.

Of course, Ecopathology is also very different from Psychopathology, and the difference reflects an important truth about Lyotardian postmodernity. The new theory is not going to emerge in the same form as its predecessors. Ecopathology is a decentred theory – it has no great source, no father figure. Whereas Psychopathology originates in a great work, Ecopathology is a diffused way of talking, and it penetrates the mainstream media in stray remarks and vivid asides. Paradoxically, as a result, Ecopathology is diffusing much faster than its psychological relative ever could. Ecopathology has both the advantages, and the disadvantages, of being an authorless theory. Nevertheless, it shows the classic features of modern enlightenment at its most radical.

Faced with a true modernist insight, one's immediate reaction is always disbelief. Surely that is an exaggeration! Ecopathology must force us into the same unwilling suspension of disbelief as Psychopathology has done. The price of seeing the truth is high. This is also where these moments of analytical extremism cause trouble for the mainstream. In the face of the petrol protests, ecology serves to relegitimate the mainstream authority of modern progress, both scientific and political. We gave up on the protestors when the rain came down. Science came to the aid of the political modernisers with their green taxes and their climate conferences. But there is also this interpretive extravagance to the ecological view: the small details of ordinary life fill with the darkest meanings. Common sense is overturned. Such Ecopathology goes far beyond the needs of the main arguments about global warming and government policy. The total effect is unstable. The authorities are drawing for support on ideas that also have this much more extremist potential, that serve to undermine sweet reasonableness and familiar common sense. Defending a new radical expertise, Beck laments

[t]*he blindness of everyday life with respect to the omnipresent, abstract, scientific threats ...*[39]

The Conclusion So Far

Foot-and-mouth

In the postmodern account, modernity is seen as a kind of forward-moving inertia – a conservatism that rolls downhill into the future. That view underestimates both the energy that goes into relegitimising mainstream modernisation and the radical strand of critique. Modernity is a far better story than postmodernism recognised. Consider the next episode, the epilogue to our millennial autumn.

In the late winter of 2000–1, the next ecological crisis expressed itself: the foot-and-mouth outbreak in the UK with its rapidly expanding global consequences. One of the first points is that everyone immediately saw this episode as a follow-up to BSE, and a part of the wider story of the corrupting effects of modern food production and consumption. In the new millennium, this crisis was instantly recognisable as an ecological one, rather than simply an issue of veterinary health.

The foot-and-mouth episode perfectly illustrates both sides of the impact of the new ecology. We can see here both 'Ecological Relegitimation' and its more radical cousin, 'Ecopathology of Everyday Life'. The same moment both reinforces and challenges the authority of mainstream politics and institutions. Foot-and-mouth displays the ambiguity of the ecological impact on the next 'wave of modernisation'.

The disease is soon tracked to its source, as this headline shows:

Disease trail led to this squalid farm
(*The Guardian*, 24 February 2001)

'Trail led' is the language of successful detection, solubility, cracking the clues. This is the new grand narrative of ecology in action: episodic; dark; and upbeat about knowledge and understanding. Foot-and-mouth has broken out. Quickly, the source has been found, and the process explained.

As the crisis deepens, so the sense of understanding grows:

The impossibly tangled web which Ministry of Agriculture staff began to unravel last week now extends across three animal species, five countries and many British counties ...
('The Making of an Epidemic', *The Guardian*, 27 February 2001)

The foot-and-mouth episode accentuates the downbeat/upbeat rhythm of our grand narrative: what an amazing discovery, what a terrible story.

The detective story starts, for the moment, when 40 sheep were sent from [a] *farm at Ponteland ...*

('The Making of an Epidemic', *The Guardian*, 27 February 2001)

The immediate impact of this crisis is not to undermine official authority. On the contrary, as the problem grows, the authority strengthens:

[T]*he full list: outbreaks and precautionary measures*

(*The Guardian*, 6 March 2001)

There is a long list of cases. But then there is almost an equal list of countermeasures. The list begins with the cracking of the clue:

Burnside Farm in Heddon-on-the-Wall, owned by brothers Ronnie and Bobby Waugh, the pig farm has been named as the probable source of the outbreak.

(*Guardian Unlimited*, 23 May 2001)

Then after the cases, there are the countermeasures:

Closed and cancelled

- *Large parts of the Lake and Peak districts are today closed to walkers*
- *Public footpaths and rights of way*
- *Crufts, Britain's biggest dog show*
- *The Forestry Commission today announced that all its forests are closed to the public*
- *Dartmoor and Exmoor National Parks*

(*The Guardian*, 6 March 2001)

Many other restrictions are anticipated. There is no serious questioning of these measures in the mainstream discussion. The 'opposition', which tried to harness the fuel protest, is reduced to querying the efficiency with which these measures are enforced.

This is the third episode in which everyday movement in Britain has been halted. First there were the petrol shortages, then the floods. Now there are the foot-and-mouth restrictions. Each time, you cannot drive down the road as you would have done. In fact, you cannot go for a walk in many places. The atmosphere surrounding the foot-and-mouth crisis shows the depth to which the new legitimacy can go. There is virtually no challenge. The political authority is confirmed by the science – we have a classic case of modernity in practice, of modern legitimation. A government that was being challenged by empty roads is now able to close off large parts of the country without any serious objections, at least for the time being.

The contrast with the petrol blockade is deep and direct:

All the cheaper cuts of meat have gone ... Stocking up – 'prudence not panic' – insisted Ellie Thornton from behind two legs of lamb ...

(*The Guardian*, 5 March 2001)

Compare this to the panic buying during the petrol blockade. Ellie Thornton is no challenge to the official order; she is just making a canny readjustment.

Far from being a moment when authority is undermined, the outbreak of foot-and-mouth underlines how deeply-rooted are the mechanisms of modern authority. When democratic politics can call on science, modern society will accept some pretty drastic actions. The foot-and-mouth crisis is also connected with other episodes:

First step reported toward early mad-cow diagnosis

Scientists in Scotland have found a first clue that may lead to a way to diagnose mad cow disease early in the infection … in order to prevent it being spread by people and animals when [they] *do not yet show symptoms.*

('Health' section, CNN.com, 1 March 2001)

Taken together, BSE and foot-and-mouth express a wider breakdown of our relations to the countryside and to the natural world. On the other hand, there is a continuous sense of advancing knowledge, new solutions around the corner. The ecological grand narrative is full of tragedy. This new modernity is, as Beck says, far darker than the previous phase. But there is also an upbeat rhythm: knowledge moves so fast, takes hold so swiftly. The big picture is almost ours:

Global warming 'can be beaten'

… They say effective technologies and measures are available …

(BBC News Online, 5 March 2001)

The crises all fit together into a big picture given by

the new ecology. Associated with this picture, there are countermeasures and remedies. They will not reverse the damage, or lift the tragedies from people. But in the terms that Lyotard dismisses in his 'Postmodern Fable': 'It is reasonable to hope'.

However, that is only one side of the story. At moments, the foot-and-mouth episode is far more unnerving. Phrases like 'out of control' also abound. This language does not really challenge the legitimacy of the authorities or their experts: on the contrary, the greater the threat, the more we must do as we are told. But still something is stirring, beneath and beyond the official version, and particularly when the gaze pauses over tiny details, insignificant scenes or incidents. Take this description of the first infected farm:

Few neighbours have a kind word to say about the collection of long concrete sheds on the western fringe of Newcastle upon Tyne that pass for a farm and are the likely source ...

(*The Guardian*, 24 February 2001)

Such images have the jolt of Ecopathology. We see in uncanny outline an everyday scene that is ugly and dirty, but in an apparently ordinary way. There must be thousands of scenes like this all over the country:

Surrounded by rusting tin drums, piles of rubbish and old machinery ...

(*The Guardian*, 24 February 2001)

But under the interpretive gaze of Ecopathology: these specific tin sheds turn out to be connected to dozens, hundreds, maybe thousands of cases of the new disease. Menace springs up behind the mundane. We listen to the neighbours moaning, as millions of neighbours will:

The place is an eyesore, a disgrace.

(*The Guardian*, 24 February 2001)

As with Psychopathology, there is no commonsense pathway leading from this ordinary surface to the deeper truth. The problem is hidden from the ordinary gaze – hidden partly by the banality of the surface. Only the Ecopathological interpreter can make the terrible connection:

... a vital link in a complex, and potentially lethal national supply chain ...

(*The Guardian*, 24 February 2001)

These are centres of hidden complexes, these everyday backyards.

Radical interpretation sweeps up the neglected scene. The expert or informed view plunges vertiginously from surface to depth. Just when you thought you had tamed Psychopathology, when it seemed safe to venture out among the trivialities again, here comes its even more threatening cousin, Ecopathology. We have just got used to living with the unconscious, we know that all our habits have connections to hidden meanings. Nothing is psychologically innocent. But now we have to begin again, and recognise that nothing is *ecologically* innocent. Of course, this is late modernism – decentred, fluid, and transmitted by the news – but it has exactly the cutting edge of the earlier phase.

What Next?

Everywhere these little explosions go off amidst the trivia. Can you trust everyday life at all? Who knows where the next harm will be discerned, in what once-cosy or reassuring place? Are those sheds just a nuisance, or are they another sleeping disaster? One of the key signs of radical modernism at work is that its interpretations outrage and bewilder their subjects. This hidden meaning comes as a shock when it is revealed, especially to those inside the picture. The owners of those old sheds are heard facing the new meaning given to their lives:

We just don't understand how it could have happened and what it means for us now. Like everyone else, we are in the dark and we are finding the whole situation very upsetting and difficult to cope with.

(*The Guardian*, 24 February 2001)

These phrases belong to all who have been reinterpreted from the perspective of a deeper theory:

'just don't understand'
'what it means'
'in the dark'
'upsetting'
'difficult to cope with'

The secret meanings are those we would like to avoid or ignore – that is the link with Psychopathology. But we would be unwise to assume that we too will not have to face our own unwanted truths and appalling interconnections. Nobody knows which of these countless ordinary scenes has the terrible secret. Maybe all of them do.

Ulrich Beck argues that in the face of ecological interpretation, people always pull back. The more revealing the insight, the more violent the recoil:

[T]*he* ***resistance*** *to insight into the threat grows with the size and proximity of the threat. The people most severely affected are often precisely the ones who deny the threat most vehemently, and they* ***must*** *deny it in order to keep on living.*[40]

'How can that be us?', lament the subjects of both Ecopathology and Psychopathology. The recoil might be called 'ecological resistance' – parallel to the psychological resistance that Freud predicted people would show in the face of his insights. Like its psychoanalytic forebear, Ecopathology always provokes this resistance from the 'common sense' of those it is subjecting to the closest analysis. Beck's comments about radical ecology at the end of the twentieth century are taken, implicitly, straight from the repertoire of radical psychology at the beginning of the century:

The perception of devastation must break through

walls of denial. ... [I]t is only rarely possible without crucial assistance from experts, that is, alternative experts ...[41]

When the focus is really close-up, Ecopathology is not an expression of common sense. On the contrary, time and again, the sensible view turns out to be exactly wrong. That drain really is a sign of a global crisis; those dirty sheds really have begun a national, even a global, plague.

Like the confidence of the Freudian Psychopathology at the start of the last century, this radical Ecopathology is also difficult to absorb. *Radical modernism is extremely hard to take.* No wonder postmodernity seemed like a good idea. Deep modernism has this unrelenting grip on all the details of life – and this way of reading more than we can take into everything, finding deep problems where we want to see only surfaces and coincidences.

Yet all the while, ecology is also a resource for legitimising the modern mainstream. Provided that the democratic authorities listen to the right advice, and allow us to hear it too, there is room to hope – and reason to modernise. That is the discourse of relegitimation. We accept that we cannot walk on those paths, or enter that whole area, because we recognise the validity of the science behind the government's measures. We are still the citizens of science-led modernity, and we are still investing in the narratives of progress. With anxiety and hope, we are still waiting for modernisation to come true.

Postmodernity is slipping into the strange history of those futures that did not materialise. Instead we will face the next instalments of modernisation, the addictive serial for the citizens of the twenty-first century. Modernisation is less predictable than postmodernity would have been. Whereas the post-modern would have been one 'condition', modernisation pulls us in different directions. Nobody can be sure which way we will go. In one direction, there is the pull of 'relegitimation', our shaken but surviving confidence in the arguments of mainstream modernisation. We need to upgrade our society – and the best chance of doing so remains the twentieth-century alliance of science and democracy. We are more aware of how that alliance keeps breaking down, how neither party can trust the other: but we are still taking a chance on this story. However, we are also subject to the repeated alarm calls of radical modernism. Every day another detail from everyday life turns out to carry a hidden meaning, a threatening subtext. We all experience daily miniature versions of the pig farmers' shock: can that be our life, out there?

In Britain's millennial autumn, the balance held. The mainstream of modernisation overcame – for now – the protests of those it cast as the new reactionaries. Meanwhile, the radical flashes went off in the margins. But in the next 'crisis', modernisation might split into its different elements – the mainstream and the radical, which we can represent, conveniently, from the US presidential election of the same autumn, by the official Democrat Al Gore and the dissident Ralph Nader. The way will open for the real 'petrol protestors', those who claim to represent the global motorist and the world's petrol economies, who will demand a real end to all environmental constraints: all those unnecessary regulations and burdens on consumers and also (they will add) on oil producers. A new alliance will stand against the modernisers, both mainstream and radical. And this time, the petrol protestors won't be standing outside the depot gates but inside the White House.

Notes

1. Gianni Vattimo, *The End of Modernity*, translated by Jon R. Snyder (Cambridge: Polity Press, 1988).
2. Ulrich Beck, *Ecological Enlightenment: Essays on the Politics of the Risk Society* [1991], translated by Mark Ritter (New Jersey: Humanities Press, 1995).
3. Ibid., p.50.
4. Jean-François Lyotard, *The Postmodern Condition: A Report on Knowledge* [1979], translated by Geoff Bennington and Brian Massumi (Manchester: Manchester University Press, 1984), p.37.
5. Ibid.
6. Ibid., p.27.
7. Milan Kundera, *The Unbearable Lightness of Being* [1984], translated by Michael Henry Heim (London: Faber & Faber, 1984), Part VI.
8. Lyotard (1984), p.37.
9. Ibid., p.40.
10. Ibid., p.51.
11. Ibid., p.60.
12. Jürgen Habermas, 'Legitimation Problems in the Modern State', from *Communication and The Evolution of Society* [1976], in William Outhwaite (ed.), *The Habermas Reader* (Cambridge: Polity Press, 1996), p.248.

13. Ibid., pp.250–1.
14. Jean-François Lyotard, *The Differend: Phrases in Dispute* [1983], translated by Georges van den Abbeele (Manchester: Manchester University Press, 1988), p.xi.
15. Ibid., p.29.
16. Ibid., p.76.
17. Ibid., p.128.
18. Jean Baudrillard, *Cool Memories* [1987], translated by Chris Turner (London: Verso, 1990), p.35.
19. Jean-François Lyotard, 'A Postmodern Fable', *Postmodern Fables* [1993], translated by Georges van den Abbeele (Minneapolis: University of Minnesota Press, 1997), p.83.
20. Ibid., p.85.
21. Ibid., p.92
22. Ibid., p.95.
23. Ibid., p.96.
24. Ibid., p.97.
25. Ibid., p.99.
26 Lyotard (1984), p.40.
27. Ibid., p.60.
28. Ulrich Beck, *Risk Society: Towards a New Modernity* [1986], translated by Mark Ritter (London: Sage Publications, 1992).
29. Lyotard (1984), p.60.

30. Ibid., p. 51.

31. Jean Baudrillard, *The Perfect Crime* [1995], translated by Chris Turner (London: Verso, 1996).

32. Sigmund Freud, *The Psychopathology of Everyday Life* [1901], translated by Alan Tyson (Harmondsworth, Middx: Penguin, 1975), p. 37.

33. Ibid.

34. Ibid., p. 38.

35. Ibid., p. 56.

36. Ibid., p. 61.

37. Ibid., p. 84.

38. Ibid., p. 94.

39. Ibid., p. 13.

40. Ulrich Beck (1995), p. 3.

41. Ibid., p. 10.

Further Reading

On postmodernity, further to the sources cited in the Notes:

Hal Foster (ed.), *Postmodern Culture* (London: Pluto Press, 1985).

Jeffrey C. Alexander and Steven Seidman, *Culture and Society: Contemporary Debates* (Cambridge: Cambridge University Press, 1990).

James Good and Irving Velody (eds), *The Politics of Postmodernity* (Cambridge: Cambridge University Press, 1998).

On the philosophers of postmodernity:

For more Lyotard, *The Inhuman*, translated by Geoffrey Bennington and Rachel Bowlby (Cambridge: Polity Press, 1991).

For more Baudrillard, *Cool Memories*, Vol. II, translated by Chris Turner (Cambridge: Polity Press, 1996), and 'Maleficent Ecology', in *The Illusion of the End*, translated by Chris Turner (Cambridge: Polity Press, 1994).

For Derrida and postmodernity, *Cinders*, translated by Ned Lukacher (Lincoln: University of Nebraska Press, 1991).

On threat and fear and postmodernity:

Ronald Collins and David Skover, *The Death of Discourse* (Colorado: Westview Press, 1996).

Elaine Showalter, *Hysteries: Hysterical Epidemics and Modern Culture* (London: Picador, 1997).

Brian Massumi (ed.), *The Politics of Everyday Fear* (Minneapolis: University of Minnesota Press, 1993).

On the philosophical roots of science, panic and hope:

Keith Ansell Pearson, *Viroid Life* (London: Routledge, 1997).

For further Beck:

Ulrich Beck and Elisabeth Beck-Gernsheim, *The Normal Chaos of Love*, translated by Mark Ritter and June Wiebel (Cambridge: Polity Press, 1995).

Further sources on new modernity:

Bruno Latour, *Aramis or The Love of Technology*, translated by Catherine Porter (Cambridge, MA: Harvard University Press, 1996).